YOUR KNOWLEDGE HAS VALUE

Airline Business. The Role of Governments in Supporting Airlines in the COVID-19 Crisis

Julia Werner

Bibliographic information published by the German National Library:

The German National Library lists this publication in the National Bibliography; detailed bibliographic data are available on the Internet at http://dnb.dnb.de.

ISBN: 9783346684615
This book is also available as an ebook.

Coursework A

The role of governments

in supporting airlines in the COVID-19 crisis

Airline Business

Due Date – 11 July 2020

Julia Werner

Word Count – 2,740

List of figures

List of abbreviations

ASA ....*Air Service Agreement*	IATA*International Air Transport Association*
BOC Aviation......*Bank of China Aviation*	
CEE......... *Central and Eastern European*	ICCA*International Congress and Convention Association*
CEO .... *Chief Executive Officer*	
COVID-19 ....*coronavirus disease 2019*	LCC.............................. *Low Cost Carrier*
EU .... *European Union*	PWC...............*PricewaterhouseCoopers*
GDP............... *Global Domestic Product*	SAF................. *Sustainable Aviation Fuel*
GmbH*"Gesellschaft mit beschränkter Haftung"*	

Figure 1 Government Bailout

1 Introduction

In this paper, the author examines the background of government intervention in the airline industry due to the grounding forced by the coronavirus disease 2019 (COVID-19). Political involvement can mean assets and drawbacks. In the end, it will be interesting how the government-airline relationship will evolve in the post-pandemic phase.

2 Background of government intervention in the airline business since the COVID-19 crisis

Out of the blue, the black swan COVID-19 circles planet earth. Closure of borders and travel bans spread like supersonic expansion. The pandemic paralyses flying principles. Within days, many airlines around the globe ground their entire fleets and run into difficulties through no faults. Why do governments intervene in the airline business?

IATA believes it is the only solution (IATA, 2020). As long as travel bans exist, it will be almost impossible for passenger airlines to connect countries and continents again.

2.1 Does government support work?

The author provides a critical evaluation of some airlines in different countries and continents. It remains questionable which airlines can make the journey through the crisis.

2.1.1 Airlines in Europe

- Norwegian Airlines has been dire straits before the pandemic. The airline could raise enough financial support from the prosperous Norwegian state. However, despite streamlining its structure, the airline had to convert debt into capital. The consequence is a 12,7 per cent share of the Chinese government in the organisation. Political entanglement is through the Bank of China, which is a shareholder of the leasing company BOC Aviationwhich in turn leases aircraft to Norwegian. The economic interdependence of a leasing company made this possible. It is questionable if future conflicts of interest allow long-term viability.

- In the case of Alitalia, the airline has filed for bankruptcy in 2017, but continued operations ever since due to financial aid by the Italian government. Since the coronavirus crisis, the state controls the airline completely. However, in February 2020, the European Commission started investigations of compliance with European Union (EU) State aid rules. A wrong decision of the Commission towards the airline could mean the final shutdown of the Italian flag carrier.

- The German leisure carrier Condor shares a similar fate. Condor received state aid before the COVID-19 crisis and was struggling to find investors. A possible solution was already signed, but the crisis turned the tide. Further state aid was required to continue operations. Viability of the organisation remains highly questionable in a restart phase which costs lots of money with limited revenue and profits.

2.1.2 Situation in Africa

In South Africa, the government cannot accept further financial burden to support South African Airways. The airline has to change its business model and corporate culture. This is the precondition that new leadership, new management and new employees can create a financially viable and competitive airline.

3 The COVID-19 situation

As outlined above, the COVID-19 lockdown has an enormous economic impact to all kind of businesses. The unique feature of the crisis is that financially sound organisations are hit likewise. In the airline industry, few organisations, e.g. cargo companies, can continue operations and business. A government bailout is for most airlines, the only option to bridge times without any revenue. Further, the cash flow burden on restart will be huge. In the best case, the first flights are cash neutral; that means that the costs of the flight are covered but without making any profit. Therefore, it is essential to analyse the chances and pitfalls of government support.

3.1 Why is government support crucial?

3.1.1 Financial support

The pandemic lockdown eliminated the business basis for airlines; thus, total revenue stopped for most airlines. Financial support stabilises the harmful cash flow situation airlines face; e.g. Lufthansa has current expenses of about 25 million euros per day and additionally open claims for refunding tickets. Without financial aid, 85 per cent of all airlines would have ceased operations due to COVID-19 burdens.

Due to copyright issues, this figure isn't included in this publication.

Figure 2 State Support for Airlines as of 19 June 2020

3.1.2 Business Continuity

The financial support guarantees business continuity. Airlines are organisations with complex systems. Business continuity is a crucial factor as a total shutdown means economic and infrastructural collapse. The continuity ensures to maintain the workforce during all phases of the crisis. The workforce is an essential resource and asset and decisive in the moment the organisation ramps up the system again.

3.1.3 Global Domestic Product (GDP)

Airlines and especially airline groups make a significant contribution to the GDP of a state. Large numbers of airline employees are above average skilled employees and contribute to the GDP to a greater extent than the average population. Therefore, maintaining the workforce current during the crisis ensures future GDP. Further, airline business generates economic multiplier effects on other industries, e.g. the travel and hospitality industry, which themselves add a large percentage to GDP.

3.1.4 Air Connectivity

Air connectivity directly impacts GDP. Connectivity means passenger movements and airlines – through their network- enable human capital flow and, business investment (PWC, 2020). Airlines in leading states offer a hub network. Austrian Airlines in Vienna or Brussels Airlines in Brussels offer primary links to different European cities like Paris, London and Mallorca in summer; but also, secondary links to destinations where only one airline offers a routing; and tertiary links which can be served by a long-haul network, e.g. Sarajevo, Pristina and Varna by Austrian Airlines.

3.1.5 Secondary effects

Airlines safeguard the economic future and growth potential of locations. The economic footprint of the aviation industry is enormous. The International Congress and Convention Association (ICCA) stated that the city of Vienna is again on the second position of the most famous congress cities globally. This fact demonstrates the significance of a hub airline, e.g.

Austrian Airlines. Further, many headquarters of Central and Eastern European (CEE) organisations are located in Vienna (Invest in Austria, 2019). Economic damage without a functioning hub airline can be devastating.

3.1.6   Sustainability

Future concepts show that post-COVID-19 operation will adopt a more sustainable way of airline operations. Airlines right-size their fleets and outsource older and less fuel-efficient aircraft, e.g. Lufthansa takes five A380s and five Boeing 747-400 aircraft out of service. As Tim Clark, CEO of Emirates, outlines that the pandemic will be the end of the Airbus A380 (Aviation Business, 2020). The aircraft is too expensive due to high running and fuel costs. In France, the government-linked sustainability conditions to the financial bailout of Air France; all flights of one-hour flight time will be suspended if the trip takes two and a half hours on a train.

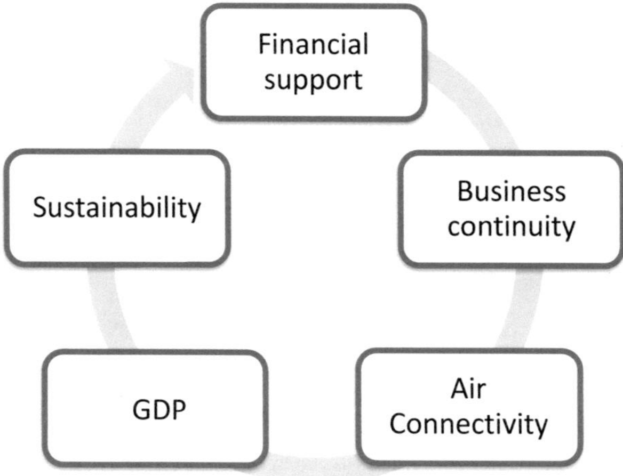

Figure 3 Why is government support crucial?

3.2   Potential pitfalls of government support

3.2.1   Government management and strings attached

Government support is linked to conditions. The state investment in stock companies is defined by law. The consequence is that state representatives will be sent to the supervisory board of airlines. However, as Lufthansa CEO Carsten Spohr (2020) points out: "we need government support, but no government management". Furthermore, as it the case in the EU, the European Commission sets further terms and restrictions on individual state aid.

3.2.2 Independence means economic freedom

Airlines like Wizz Air and Ryanair are almost independent of any government ties. This fact offers a competitive advantage for the years following the pandemic. The organisations maintain their full economic flexibility and can adopt a business as needed. Further, the low-cost carriers (LCC) will strictly focus on their business model as Wizz Air's CEO Jozsef Varadi pointed out in a webcast held by World Aviation Festival on May 26, 2020.

3.2.3 Distortion of competition

The Pre-COVID-19 situation was that overcapacities exist in particular in the European aviation market. Market adjustments are necessary now more than ever. State aid distorts market shares and slows down the downsizing of capacities. Companies which do not seek state support see themselves disadvantaged.

3.2.4 Burden of debt

The burden of debt is enormous, and the repayments will raise new challenges for many airlines. For organisations, it is essential to prevent a debt overload situation. This could soon be the case if companies cannot present a solid business plan. Also, repayments of state aid will be the first objective when companies start to make a profit again. Organisations have to limit or postpone investments which are necessary to stay competitive.

3.2.5 Viable airlines

Not all airlines which receive state aid are viable independent of the COVID-19 crisis; e.g. Condor Flugdienst GmbH receives state aid for the second time within 12 months, and it is questionable if the airline like many others will be viable after the restart of business in long-term or if the financial support will prolong the dire strait of the organisation. It is evident that many airlines - which have been struggling for years – will go bust in the aftermath of the crisis. A further focus will be on the viability of Alitalia and Norwegian Airlines.

3.2.6 Reversal of the privatisation

Collaboration in the past did not work. E.g. Lufthansa was fully privatised in 1997. Now, the German state will be initially a silent partner but has the option to convert the silent participation in share value of the 25 per cent plus one share to protect the organisation from foreign ownership and interest payment obligations. However, liberalisation is critical for the future success of airlines.

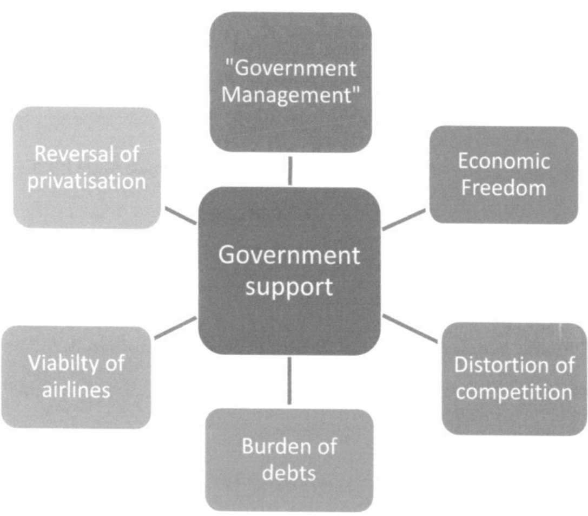

Figure 4 Potential pitfalls of government support

4 The government-airline relationship of the future

4.1 Is a fresh start needed?

The future government-airline relationship will undoubtedly differ from the pre-COVID-19 time. The growing gap in the development of government and airline interests can get a chance for a reset at a common ground. A stable government will back up a robust airline industry and, vice versa, a robust airline network can support states' social and economic interests. The crisis revealed the interdependence in a particular way. The pandemic has brought to light a specific situation in Europe. The EU has missed being united in the crisis; the union disappeared, and, coordination went out of control. Every country takes different measures; thus, the pandemic is much more harmful to the economy. The future objective must be that the EU as a system works.

4.2 New Normal

4.2.1 Government-airline relationship

What will be the new normal in the government-airline relationship? Indeed, the primary goal will not be a reversal of airlines' privatisation. However, the relationship between both parties will be more intense than before the crisis. The post-COVID-19 time can be an opportunity for more collaboration and more intense regular exchange of interests. For the industry to ramp up again, governments are challenged to define new regulations, a legal framework regarding changed hygiene and health requirement to facilitate business continuity with pandemic preparedness and contingency plan. The task of the airline industry is to develop a new strategy based on the health and safety regulations defined by governments. Although, standard social distancing requirements will make most airlines unviable if the load factor is artificially capped by 62 per cent (IATA, 2020). Challenges in the ramp-up will be the time. Legislation usually is a prolonged process, but after the corona crisis, the continuity of airlines will be driven by the pace how fast governments will arrange guidelines and organisations can be back in the market before any competitor fills in the demand. Moreover, corporate sustainability must be assured.

4.2.2 Europe

Europe has missed being united in this crisis. Every country is doing things differently. Some even say the EU has disappeared, and the system went out of control. Thus, the crisis is much more harmful to the economy. In future, the EU must be better coordinated and do a better job as a system. The pandemic must be a learning experience. It is crucial that there are no separate approaches, no national solos, in legislation and taxation.

4.2.3 Domestic traffic

Domestic traffic within China and later within the USA might function best due to the size of the market. The restart and growth of intra-European traffic will depend on how to open member countries are in solidarity or if all measurements and procedures are served as a piecemeal. LCCs will be more flexible than legacy carriers. An advantage will be that LCC is not highly dependent on business travellers. The market share of LCC before the pandemic was 45 per cent within Europe; post-crisis it will rise towards 50 per cent within Europe.

4.2.4 Government strategy

Development of new strategies of governments is a core solution. If one compares the black swan with a green swan, the main difference is that green swan events are far more destructive and require a radical new policy approach (Pareira da Silva, 2020). The effects are more direct and massive, with a high degree of complexity. That also means that policies are less restrictive but more supportive of innovative and viable airlines in the future. Regulation is needed to support future growth, but not to limit it.

An example is a French state which intervenes in domestic traffic of Air France. However, ties are not the best for new investments; Liberalisation can assure new routes and traffic growth. Differentiability of markets is crucial; there are countries with a higher catalytic impact of aviation like Greece and countries with a different growth potential of air activity.

Essential progress is needed in data mining, new legislation about sharing and storing data, possessing and untangling data. Furthermore, how do we get external data? The author sees a crucial task in radically identifying new ways of how to convince potential passengers to be willing to share data. Present studies show that participants who do not trust the government are less likely to install a corona warn app.

The following graph shows that people who trust the government are more likely to install the app, whereas participants with little trust in governments are not willing to install the app and make use of the new technology.

Participants who don't trust the government are less likely to install.

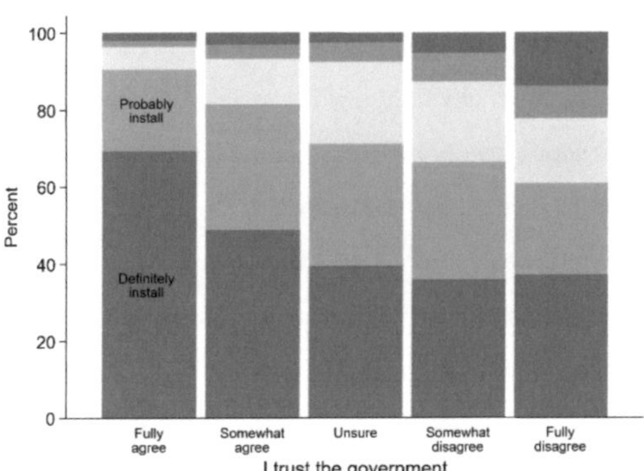

Figure 5 Trust in government is essential

Further, it is observed that the EU is opening the intra-European market but struggles with granting access to markets like Turkey. Cause for this are economic reasons, to keep the money within the EU, but also differences in coronavirus handling procedures. This example demonstrates the challenge of negotiating timely bilateral agreements. Indeed, the ramp-up situation is challenging as many businesses are still shut down, and governments work together to protect countries and its population from COVID-19; e.g. foreign travel is difficult because of lasting quarantine rules. Air Service agreements (ASA) are newly negotiated; long-term solutions are required in case of future uncertainties like a second wave hitting partly or extensively the aviation market. A principal aim must be a standard lockdown procedure; among other things, including solutions for banned free travel, support in testing, availability of results and isolation procedures.

A new approach could also be new partnerships; e.g. the government of Ruanda and Qatar Airways partner to build a new airport and new capabilities.

The further government needs to support innovative technologies for sustainable measures, and partner with airlines in manufacturing and usage of sustainable aviation fuel (SAF); governments require to regulate the production of SAF.

In every aspect, the author emphasises that collaboration, cooperation, and consistency are vital for all stakeholders, also to guarantee the future success of the EU market. The Distinctive approaches during the crisis need to be adopted as a new normal. Close monitoring of the industry will ensure to secure locational advantages and the attractiveness of airlines for managers, employees and not least for shareholders.

5 Conclusion

The author concludes that in the COVID-19 situation, government support is vital for various reasons. The aid is not a choice but a lifeline for airlines. Financial support ensures business continuity. The airline industry, in return, offers connectivity to ensure an immense contribution to GDP and economic wealth of countries. Sustainability in aviation plays an important role, also in the post-COVID-19 world.

However, there are some potential pitfalls and strings seen in the context of government subsidising airlines. It is essential that airlines are not renationalised and remain economically independent; the latter can be decisive for fair competition. Subsidised airlines might struggle under the burden of debt, and in the years to come, their viability depends on future business opportunities and restructuring.

Regarding the future relationship between governments and airlines, new strategies are required; a radical new mindset is vital. It might be a first for governments and airlines to cooperate closer than ever before to ensure a sustainable aviation infrastructure and national economic strength.

6 References

6.1 Journals

Kerr, S. (2020), *"France prepares to nationalise in coronavirus war"*, Global Capital, London

The editorial board (2020), *"Airlines should not get blank cheques to survive"*, FT.com

Briancon, P. (2020), *"Nationalizations, bailouts, state guarantees: Why big government is making a comeback"*, Barron's

Weblog Post (2020), *"The cranky flier: 3 links I love: Alitalia gets great news, contour sneaks up, how planes fly"*, Newstex Trade &Industry Blogs, Chatham: Newstex

6.2 Webpages

IATA (2020), *Social distancing would make most airlines financially unviable*, available at: https://www.iata.org/en/iata-repository/publications/economic-reports/social-distancing-would-make-most-airlines-financially-unviable/ (Accessed: 18 May,2020)

Aeronews (2020), *What aviation will look like post-COVID-19*, available at: https://www-aeronewsx-com.cdn.ampproject.org/c/s/www.aeronewsx.com/amp/what-aviation-will-look-like-post-covid-19 (Accessed: 18 May 2020)

CNBC (2020), *When it is safe to fly and what will flying be like after* coronavirus, available at: https://www.cnbc.com/2020/05/18/when-is-it-safe-to-fly-and-what-will-flying-be-like-after-coronavirus.html?source=sharebar%7Clinkedin&par=sharebar&fbclid=IwAR21a5JjhCwEXUZYV093gHxbWAJ20vyZWv3L133XmthFJP1Mqn0Xw0p81gM (Accessed: 18 May 2020)

PWC, *Air connectivity: Why it matters and how to support growth*, available at: https://www.pwc.com/gx/en/capital-projects-infrastructure/pdf/pwc-air-connectivity.pdf (Accessed: 26 May 2020)

Financial Times (2015), *Companies of the east use Vienna as a springboard to the west*, available at: https://www.ft.com/content/edd57e3a-669a-11e5-97d0-1456a776a4f5 (Accessed: 26 May 2020)

Invest in Austria (2019), *Vienna continues to be popular as a congress location*, available at: https://investinaustria.at/en/news/2019/05/vienna-as-a-congress-venue.php (Accessed: 26 May 2020)

CAPA (2020), *SAA government and unions to pursue a new financially viable and competitive airline*, available at: https://centreforaviation.com/news/saa-government-and-unions-to-pursue-new-financially-viable-and-competitive-airline-992897 (Accessed: 26 May 2020)

Beresnevicius, R. (2020), *Will coronavirus be the culprit to put Alitalia to rest?*, available at: https://www.aerotime.aero/rytis.beresnevicius/24628-coronavirus-impact-alitalia (Accessed: 26 May 2020)

Aviation Business (2020), *Emirates' Tim Clark: The Airbus A380 and Boeing 747 are over*, available at: https://www.aviationbusinessme.com/airlines/21542-emirates-sir-tim-clark-the-airbus-a380-and-boeing-747-are-over (Accessed: 26 May 2020)

Breakingnews (2020), *Ryanair leads travel stock surge as holiday hotspots prepare to reopen*, available at: https://www.breakingnews.ie/business/ryanair-leads-travel-stock-surge-as-holiday-hotspots-prepare-to-reopen-1001843.html (Accessed: 26 May 2020)

BOC Aviation (2020), *About Us*, available at: https://www.bocaviation.com/en/About-Us (Accessed: 27 May 2020)

Harrell, E. (2020), *Looking to the future of air travel*, available at: https://hbr-org.cdn.ampproject.org/c/s/hbr.org/amp/2020/05/looking-to-the-future-of-air-travel (Accessed: 06 June 2020)

Müller, H. (2020), *Wer hat Angst vorm grünen Schwan?*, available at: https://www.manager-magazin.de/finanzen/boerse/gruener-schwan-coronavirus-leitet-zeit-der-grosskrisen-ein-a-1307041.html (Accessed: 06 June 2020)

Khadem, N., Chalmers, S. (2020), *Coronavirus has Virgin seeking government help and Qantas up in arms. What could a bailout look like?*, Available at: https://www.abc.net.au/news/2020-04-02/coronavirus-airline-bailout-virgin-qantas-failure/12110064 (Accessed: 27 June 2020)

Mariz, E. (2020), *Covid-19 bailouts: Lufthansa bailout in question, countries mull new aid*, available at: https://www.ishkaglobal.com/News/Article/6228/Covid-19-bailouts-Lufthansa-bailout-in-question-countries-mull-new-aid (Accessed: 27 June 2020)

Pereira da Silva, L.-A. (2020), *Luiz-Awazu Pereira da Silva, Deputy General of the Bank of International Settlements*, available at: https://www.globalcapital.com/article/b1m2j1y8y93flj/luiz-awazu-pereira-da-silva-deputy-general-of-the-bank-for-international-settlements (Accessed: 30 June 2020)